5th Grade Math

Volume 4

© 2013 OnBoard Academics, Inc
Newburyport, MA 01950
800-596-3175

---

www.onboardacademics.com
ISBN: 978-1494857233

Table of Contents

Factors and Multiples

Key Vocabulary

factor

multiple

LCM

GCF

prime number

Study the figures below to discover which ones represents the factors of 16 and which ones don't.

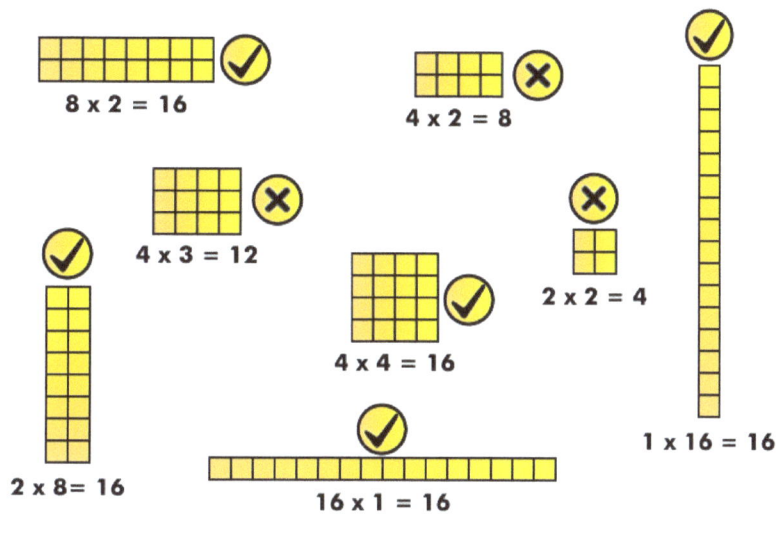

8 x 2 = 16 ✓

4 x 2 = 8 ✗

✓ 1 x 16 = 16

4 x 3 = 12 ✗

✗ 2 x 2 = 4

4 x 4 = 16 ✓

✓ 2 x 8 = 16

16 x 1 = 16 ✓

What are the factors of 16?

Fill in the empty boxes with your answers.

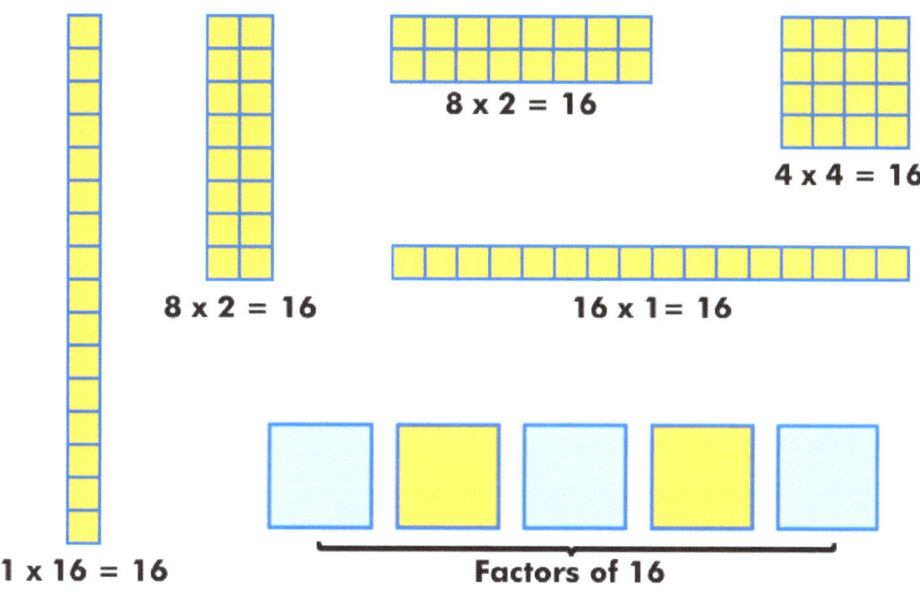

8 x 2 = 16

4 x 4 = 16

8 x 2 = 16   16 x 1 = 16

1 x 16 = 16   Factors of 16

Sort the factor pairs into the proper blue boxes.
Notice if there is anything peculiar about any of the numbers.

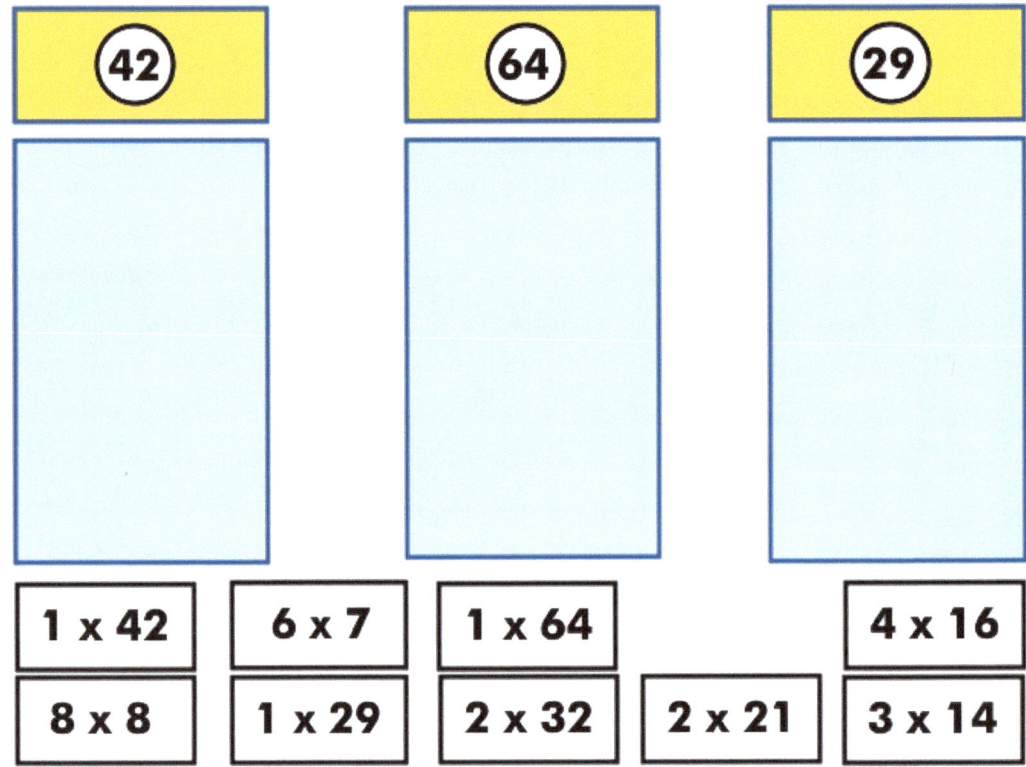

What do you notice about 29?

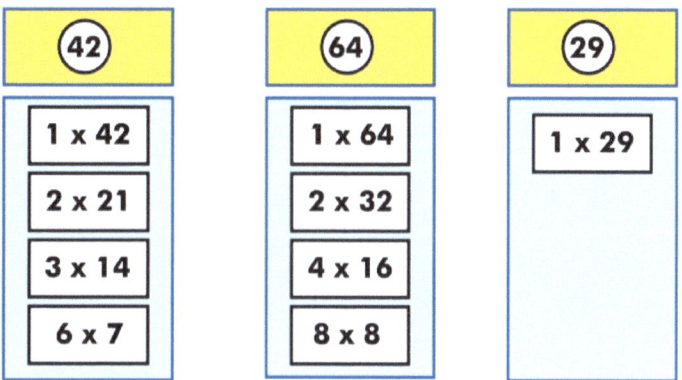

What do call a postive integer that has exactly two positive integer factors: one and the number itself?

A prime number.

_____

Sort the factors.
Write the factors for 32 in the blue, the factors for 24 in the yellow and if the number is a factor for both numbers, write it in the green.

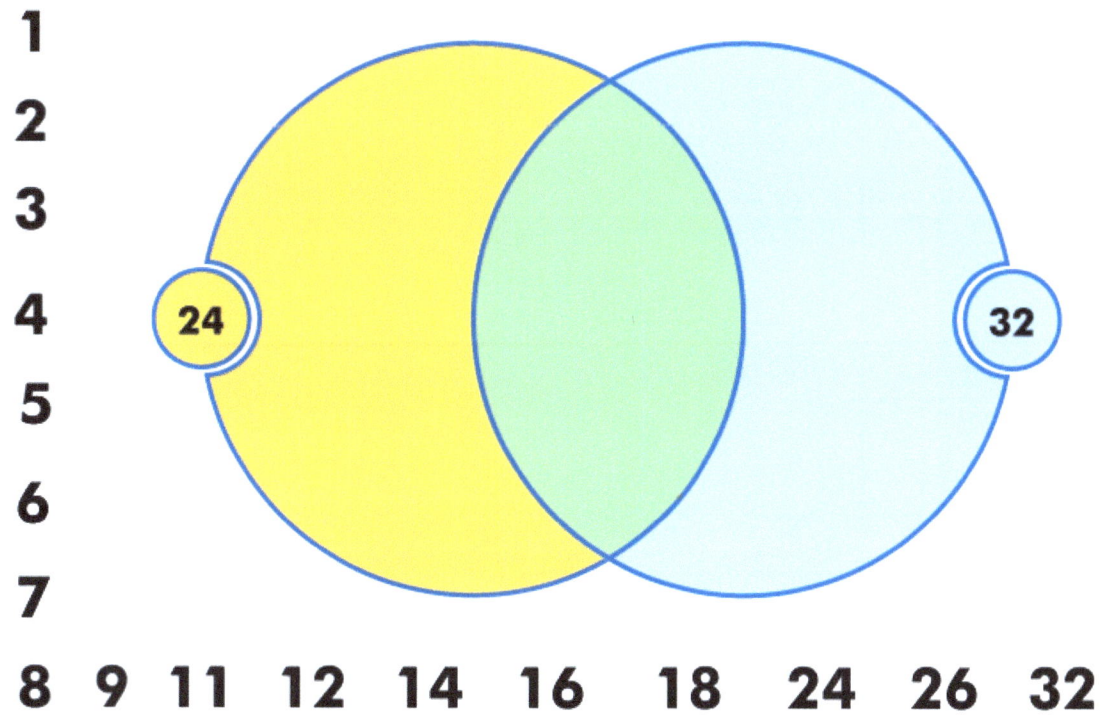

1
2
3
4
5
6
7
8 9 11 12 14 16 18 24 26 32

Circle the greatest common factor (GCF) of 24 and 32.

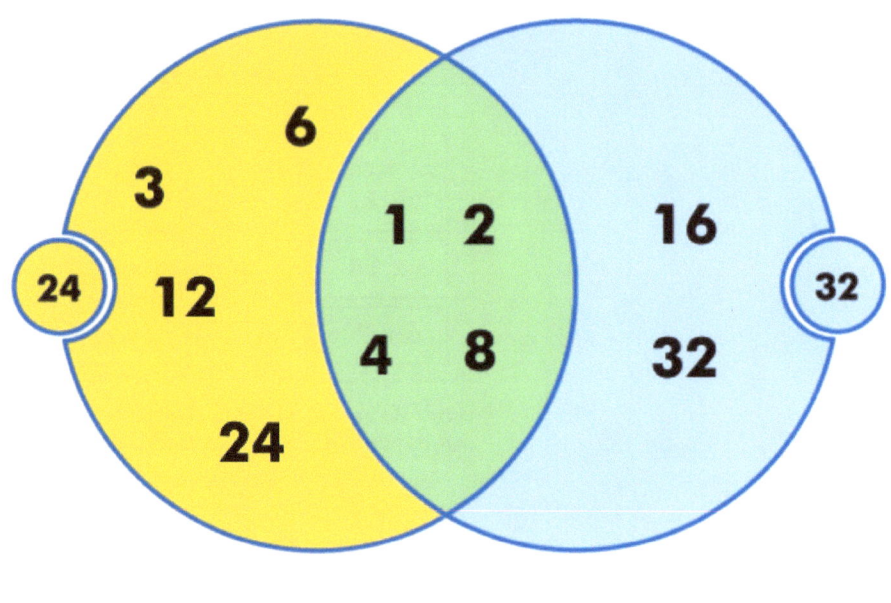

_____

Multiples and Common Multiples
(The multiples for 8 are completed to start you off.)

The first five multiples of 8:

8	16	24	32	40

The first six multiples of 4:

The first three multiples of 12:

Which number is common to all three lists?

What do we call this number?   **A Common Multiple. 24 is also the Least Common Multiple (LCM) of 8, 4 and 12.**

Find the multiples of 6 and the multiples of 4.
Draw a circle on the top for multiples of 6 and on the bottom for multiples of 4.

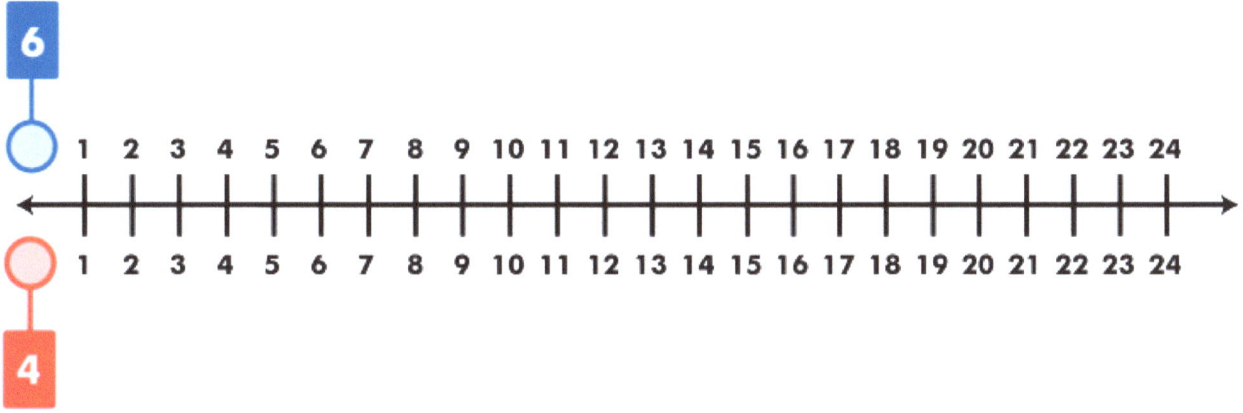

What are the common multiples of 4 and 6?

What is the least common multiple (LCM) of 4 and 6?

Find the GCF and LCM for 12 and for 20.

Factors of 12

Factors of 20

GCF

First six multiples of 12

First six multiples of 20

LCM

Name_____

Factors and Multiples Quiz

1 What is the GCF of 48 and 60?

2 What is the LCM of 17 and 51?

A 3

B 17

C 51

D 867

3 What is the GCF of 12, 15 and 36?

4 What is the LCM of 6, 7 and 14?

Integers

Key Vocabulary

integer

positive integer

negative integer

Compare these integers.

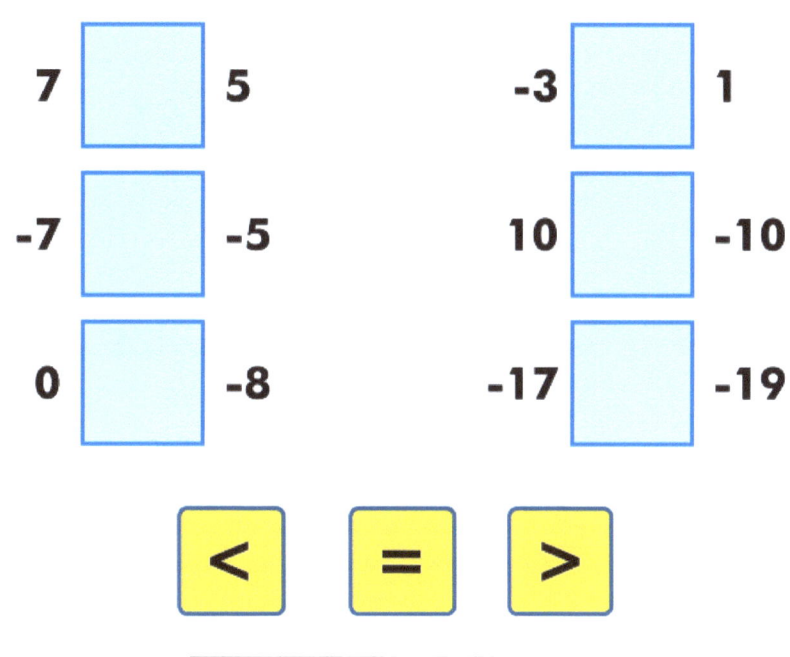

Integers on a Number Line

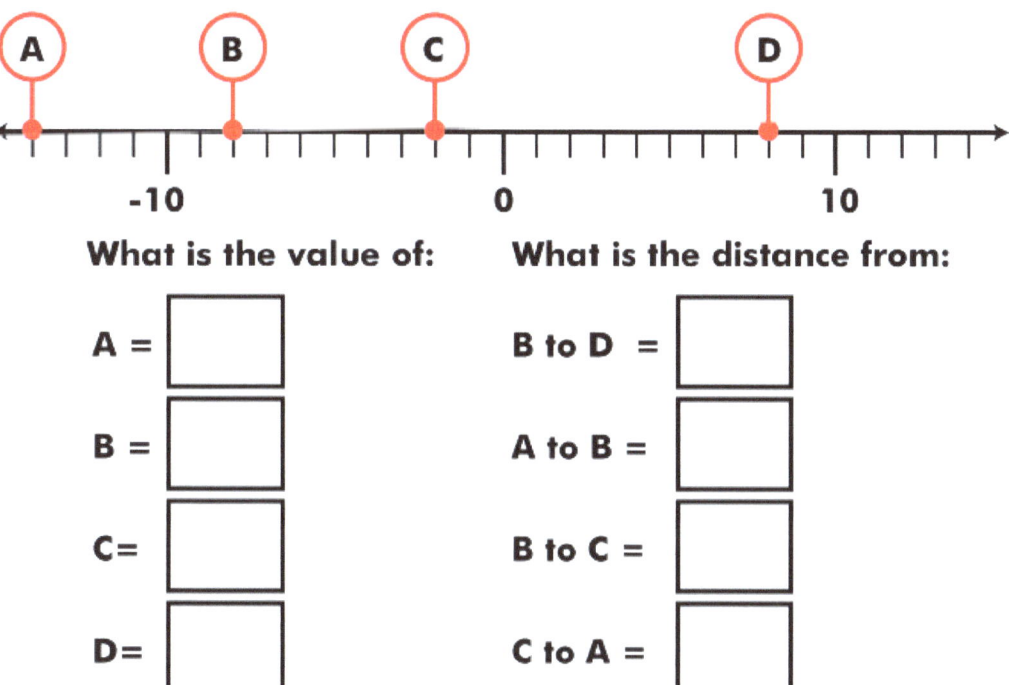

What is the value of:

A =

B =

C=

D=

What is the distance from:

B to D =

A to B =

B to C =

C to A =

Find the temperature in Winnipeg.

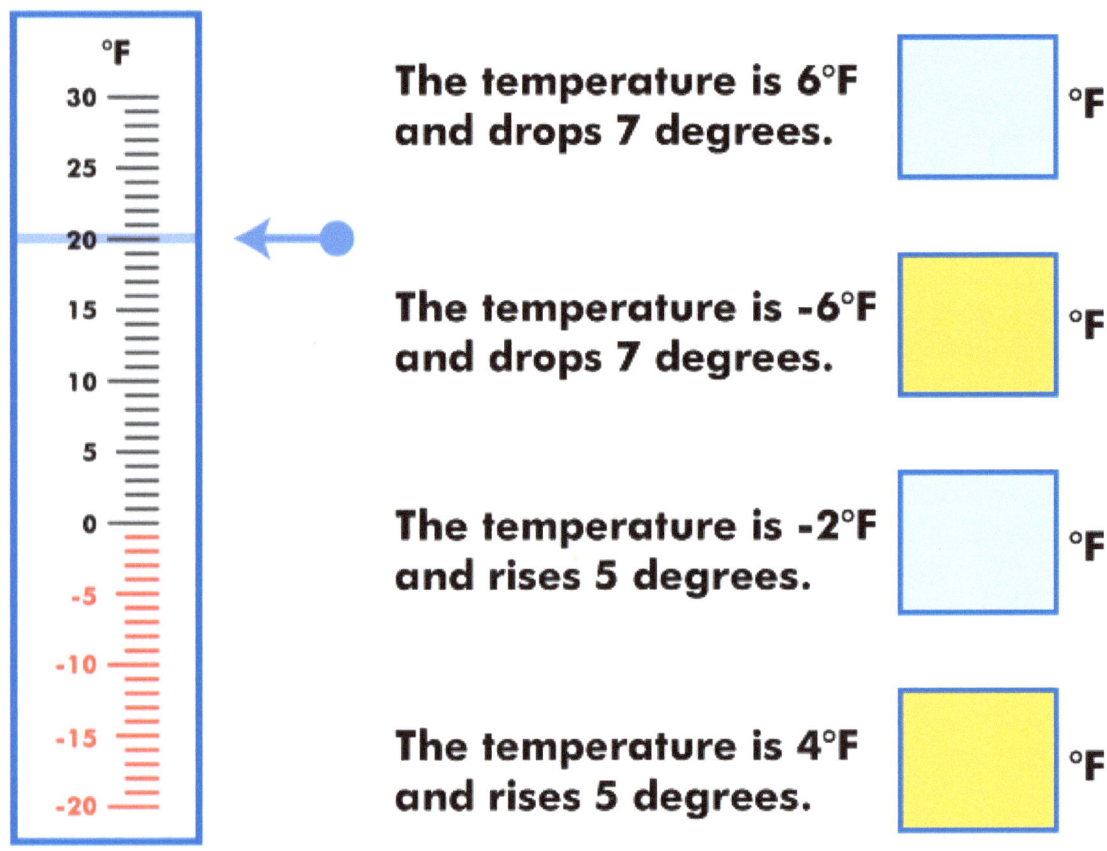

The temperature is 6°F and drops 7 degrees.   °F

The temperature is -6°F and drops 7 degrees.   °F

The temperature is -2°F and rises 5 degrees.   °F

The temperature is 4°F and rises 5 degrees.   °F

Adding Integers Using a Chip Board

$$-2 + 4 = \boxed{?}$$

Step 1: Draw the proper number of negative chips for the first term.

Step 2: Draw the proper number of positive chips for the second term.

Step 3: Simplify by crossing off the same amount of negative and positive chips.

Subtracting integers using a chip board.

$$3 - (-4) = \boxed{?}$$

Step 1: Draw the chips to represent the first term (3 + chips).

Step 2: In order to represent the second term, draw and equal number of + and - chips (4 + chips and 4 - chips)

Step 3: Model the second term by removing 4 - chips.

Practice using a chip board.

Name: _____

Integers Quiz

1 True or false? -4 > -3

2 Which integers are not ordered correctly (least to greatest)?

A -15, -11, -9, -3, -1, 0

B -3, 1, 15, 18, 22

C -5, 1, 0, 4, 7, 10

D -8, -7, -5, -4, -3, 12, 24

3 -2 + 6 = ?

4 5 − -2 = ?

Patterns & Functions

Key Vocabulary

variable

function

function table

What is the rule?

Complete the table.

Figure	Stigmas	Petals
1	1	8
2	2	16
3	3	24
4	4	
5	5	
10	10	

The rule

To find the number of petals, multiply the number of stigmas by 8.

Using Modeling
Model the seating for two and three tables and then complete the table and find the rule.
A hint is provided.

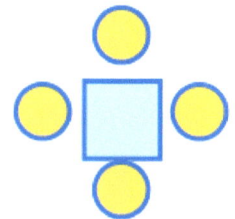  

Tables	Chairs
1	4
2	
3	
4	
5	
100	

1 x 2 + 2

Draw figure 4, complete the table and state the rule.

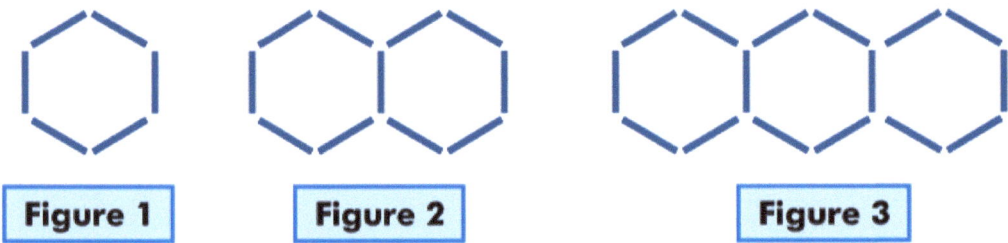

| Figure 1 | | Figure 2 | | | Figure 3 |

Figure 4

What is the rule?

_____

Figure	Sticks
1	6
2	11
3	16
4	21
5	
100	

Write the rule as an equation using variables.
Study the second example as it has been completed for you.

x	y
Stigmas	**Petals**
1	8
2	16
3	24
4	32

Equation

x	y
Tables	**Chairs**
1	6
2	10
3	14
4	18

Equation

$y = 4x + 2$

x	y
Figure	**Sticks**
1	4
2	6
3	8
4	10

Equation

Complete these function tables.

$y = 2x - 2$

x	y
3	4
5	8
7	
9	
11	

$y = x + 4$

x	y
1	
	6
3	
	9
10	

Find the rule, write the equation and complete the table.

x	y
2	3
3	4
4	5
5	6
100	

y =

x	y
1	4
2	9
3	14
4	19
100	

y =

Write your own personal equation and complete the table.

$$y = \boxed{}$$

x	y
1	
2	
3	
4	

Name: _____

Patterns & Function Quiz

1 True or false? In a function, the y value is always greater than the x value.

2 What is the missing value in this function table?

A 31
B 29
C 14
D 16

$y = 3x - 1$

x	y
1	2
2	5
10	

3 How many squares are in the next sequence?

4 If $y = 4x$, what is the value of x when $y = 48$?

Algebraic Expressions

Key Vocabulary

Expression

variable

constant

operator

evaluate

Sort the words to match the symbols.

| + Add | − Subtract | ÷ Divide | x Multiply |

minus	twice	times	plus
double	sum	more than	less than
share	increase	decrease	difference

Write the expression in terms of *n* for the Facebook friends.

Fernando has n Facebook Friends.

Mia has 10 more Facebook Friends than Fernando.

Alison has 14 less Facebook Friends than Fernando.

Mia's **facebook** Friends

Alison's **facebook** Friends

Write an expression in terms of *n*.

1   6 more than a number

2   A number plus 4

3   12 minus a number

4   A number shared between 3

5   3 shared between a number

Write the expressions in terms of _y_.

1. Last year the price of gas was y dollars. It's now 2 dollars cheaper.

2. Henry weighed y lb when he was born. He's now 4 times heavier.

3. Tori's younger sister gets y dollars allowance. Tori gets $2.50 more.

4. Write a story for the expression $\frac{y}{5}$.

Evaluate these expressions.
The first one is done for you as an example.

$a = \boxed{2}$ $b = \boxed{3}$

| 1 | $a + 5 =$ | $2 + 5 = 7$ |

| 2 | $4b =$ | |

| 3 | $4b - 2a =$ | |

| 4 | $ab =$ | |

Name: _____

Algebraic Expressions Quiz

1   True or false? $n - 5 = 5 - n$

2   There are 24 cookies on the plate. Jenna and Torri eat n cookies each. How many cookies are left?

    A $2n$

    B $12n$

    C $24 + 2n$

    D $24 - 2n$

3   Evaluate the expression $2x + 3y$ if $x = 2$ and $y = 4$

4   Evaluate the expression x^2 if $x = 2$

Newburyport, MA 01950

1-800-596-3175

OnBoard Academics employs teachers to make lessons for teachers! We create and publish a wide range of aligned lessons in math, science and ELA for use on most EdTech devices including whiteboard, tablets, computers and pdfs for printing.

All of our lessons are aligned to the common core, the Next Generation Science Standards and all state standards.

If you like our products please visit our website for information on individual lessons, teachers licenses, building licenses, district licenses and subscriptions.

Thank you for using OnBoard Academic products.